Wisdom
By
Ella

Preface:

It was the summer of 1980 when I met Muffy at the library of Southern Utah State College located in Cedar City.

I was a foreign student from Iran and had difficulties with the English language. Muffy was tutoring English at the college. As soon as I met Muffy, I realized there was something special about her. It wasn't just because she was considered legally blind and had physical disabilities, but there was something unique about her. I decided to meet with her at her home a few times. The more we grew to know one another, she slowly mentored me and revealed her knowledge about her near-death experiences.

I was very honored and pleased that she had trusted me enough to share something that she hadn't shared with others.

The main script that you are about to read is exciting and insightful. The Light of our Creator emanates from her in both this world and the next.

I Hope you enjoy this book as much as I did when I heard it directly from the Lady of Wisdom.

Syrus Saifizadeh

THE LADY OF WISDOM APPEARS

The information you are about to see, hear, or read should be put on tape, disc, or any other method you have for keeping the material so you can re-see it, re-read it, re-hear it with friends and family for lively discussions. It is that important. It is entirely new for you but it is very, very old, being assembled when the planet was in the planning stage. It is in the field of science mostly. It is about things for which you have had little or no information over the ages.

This material is for every race of people, every nationality, every age, remember that wherever you are. That is why it is coming to you over the Internet in the beginning, then other ways more familiar to you. The Internet makes sure the information goes around the world instantly.

The message is coming to you in English, with Important names in Spanish because of its higher pitch and musical sounds. In English the god sound is made with a throat clearing noise. That isn't pretty. Try it several times and listen to yourself. So, every time a god sound is used, we are translating it to El Dio

LasDiosas, a musical sound, which alone is El Dio, La Dia, or EL and LA. The complete godhead, the THEY of Genesis I verse 26-27-28.

to El Dio is of the Universal Mind, the Creator, and La Dia is Light, as the Day. Which creates life in every creature and plant as well as human beings. It is LOVE-not the romantic type, but that of friends,

Children, other people you care about and want to help, to heal, to comfort. This is the first bit of information.

Many people think it is religion. It is a fact of life, without which there would be no planet or life.

For those of you who have never met the Great Lady, nor ever realized there was one, turn to

Revelations 12, verses 1 and 2 in the New Testament of the Bible. The 12 Stars in her crown are souls, just as you also are part of a circle of 12 souls, with one greater soul in the center. That pattern of number 12 is all over the planet in many ways – 4 of these circles

makes the number 52-a sacred one.

So, the second thing you are learning is that all people are part of a circle of 12. Not just the old gods:

Zeus, Jupiter, and Odin, with 12 children each. Jesus had 12 apostles, Jacob and Esau with 12 children each. Also, Mohamed had 12 prophets.

Part of your circle may be in your home, your neighborhood, your country, but parts of it may be across the ocean in a land you may feel drawn to.

Where you are likely to go in a war, to serve others, to travel, and there meet you mate, your best friend, one who saves your life, or in war, even kills you.

Who is speaking? A soul who has played the part of WISDOM over the ages, by different names in different areas: Sophia in Old Hebrew Tara the Enlightened One in Tibetan Buddhism and Athena in ancient Greece. Now you will remember her as ELLA, EL-LA, who has the two parts of Dio-Dia in balance and is very old and wise. Ella is Spanish for SHE. And She talks in the SCIENCE field today. A woman in science? Oh, yes.

What part of science is she teaching you? About The total atomic element chart and its seven divisions, with a clue to the one that teaches you to levitate yourself and very heavy objects, such as the blocks in the Egyptian pyramids. Without machinery or other material things. She will show you the current star map and the muddled pictures of it, then tell you about four large maps and show you how to make them and then she will show you the twelve-star maps in 30-degree locations encircling the globe.

Roman poets changed the star maps 2,000 years ago to honor Julius Caesar, and his nephew Octavia, who changed his name to Augustus (July and August). Caesar was honored because he was a

great warrior who took over many lands from the people who were born there, including the British Isles. Latin was his

Language. But when he was 44, he was killed for being too big for his britches.

Star events in the sky had a lot to do with his being honored as a god, though they were meant for a spiritual El Dio LasDiosas called Jesus by English speaking people and ISSA in other lands.

Another reason for using some Spanish words is because it has a simple family of five vowels, the Latin ones. ÃÊÎÖU

English has three families of five vowels, each different from the others: Gaelic ÃEÎÖU, and Anglo-Saxon, ÃEÎÖU plus the Latin ones ÃEÎÖU.

The simple sound and spelling make it easier to learn. In addition, English has as many as eight different meanings for some sounds as words, with

Different spellings so if you hear the sound and visualize the wrong meaning, the message will be misunderstood. Besides, you can whistle Spanish and send a grocery list from the mountain, across the valley to the seashore, as is done in the Canary Islands.

The Great Mother will be honored in this New Age for she has traveled all over the world, appearing in small villages everywhere. She is telling one person, or crowds of people what is about to happen to the Earth. A great Light is soon to appear and change a familiar thing. Your mind will be so amazed you will think differently. Remember that Moses was an astronomer. Since he grew up in the royal family of Pharaohs in Egypt, and he knew what was about to Happen in the sky and that it would be a great time to free the slaves who had been in Egypt for 400 years. America, does that sound familiar to you?

The event Moses foresaw was probably a comet, seen for 52 years, witnessed by all other lands too. A book by Emmanuel Velikovsky, WORLDS IN COLLISION describes what happened more thoroughly than any other researcher. He also wrote a book about how the amnesia cultures go through denial of great catastrophes.

The Great Mother has appeared over 22,000 times in the past 2,000 years. Always coming to warn of a

great negative event, such as the potato famine in the 1800s that starved millions and sent millions to the United States, Canada, and Mexico. Videos in Catholic book and video stores tell us of her many appearances all over the world. Beginning in 1968 in Zeitoun, near Cairo, Egypt. There she appeared for years over a Coptic Christian church nightly, and

healed people with terminal illnesses – no matter what religion or if they had none at all. Then in 1981 she appeared in lands the surprise non-Catholics: Japan, lands in South America, European lands, Russia, as well as many others.

St. Thomas, the doubter of Jesus' apostles, converted the Coptic Egyptians 2,000 years ago, 300 years before the Bible was compiled by men from all over the Mediterranean, and 300 years before Rome began to accept Christianity.

Now you will see the atomic chart which is 9x12 squares on a piece of typing paper, such as the Wisdom used for the one you see here. Each of the 108 elements will have a square with its name on it and it's code in the top right-hand comer. Your number 9 squares down the page, and move over to the one next to No. 9, now in reverse go up the page to 18, next to 18 is 19, go down the page and do this manner of numbering to the end, which is at the right-hand top.

The number 108, Hy-8, is sacred in Tibet.

You need the numbers and names to help you do the color patterns within the chart. We will start with the red division, which begins with element 93. As you see on the screen, you form a shape with elements, which looks like an American Indian figure drawn on rock cliffs.

Next you go to the center and make the same figure in yellow for the heavy earth materials. As you see on the screen.

The third one to the left will be the blue area, make the same shape. With blue to the left, yellow in the center one, and red in the danger zone, it is easy to see the elements and you may know what they are. But notice that the space around each of these three also has a pattern, in reverse to what you just made. That contains three figures too. Now you have six. The seventh one is the strip across the top, which touches each strip of elements. Now, in one of these divisions is the secret for levitation, of yourself and heavy objects with no need of machinery or aids. All over the world are pyramids built by people who used to know the formula.

And when you know it too you also can fly without wings like the characters on children's early morning television.

It is possible for some teenagers as young as 14 to discover the hidden formula in the atomic chart and tell the whole world on the Internet. But you don't even have to be educated or be more than very poor to discover a great truth. Philo Farnsworth was 14 when He discovered the formulas for the television tube. He was very poor and was born on a tiny farm beside what is now interstate 15, in Southern Utah – beside the off ramp for a tiny town called Paragonah. He grew up in Idaho. In the 1980s, a teenage boy, under educated, working in a bean field in Africa's tiniest area, Kibeho, in Rwanda had his life changed when The Master Teacher came and worked beside him in the field and told him what he was meant to do. Now the boy is a grown man and serving others faithfully. There is a video about the visit. You can find it in Catholic video stores. It is called Kibeho.

All of you teenagers are meant to be discoverers of great truth, teachers and examples of them. You have grown out of your memory of who you were in the life before this one, what your assignment was when you were born. But with prayer, talking to and listening to your god, your angels, guides, teachers, you will learn where you are meant to serve others. That is the way you lift yourself, by helping lift somebody else.

Everyone is sent into mortal life and body to lift somebody who would fall more deeply without your knowledge and consideration. You have heard the message before: Love your neighbor as yourself.

Everybody is your neighbor. Some of your neighbors may be from your own spiritual circle of 12 and need your help to get back on track, without preaching, just by being an example to them.

Did you notice anything familiar in the atomic

Chart. It has the same pattern that the farmer makes in plowing. The same pattern men and women make in weaving cloth or carpets. The Persians have woven 9×12 carpets for thousands of years. Another thing the ancient Persians did was write 9×12 patterns on mud tablets.

When modern archaeologists tried to read the mud writing they thought there were two languages on the tablets. It took a while for them to find out it was all Persian, just that every other line went in the opposite direction.

Another law about reversal is that of reflections on a mirror or in water or any shiny surface.

Sometimes in your dreams you see a reversal of the sex of a character of a color, or who does an action, but after you think about it for a while you will see what it is telling you. There is also a reversal in your Language. The same message can be read either backwards or forwards. El Dio LasDiosas put clues in that manner.

In music, on a piano keyboard there is the number 12 made by number seven white keys and five black

keys, in the key of C. If you are a natural mathematician, you will find numbers in much of music and perhaps discover something there no one knew before.

There is the number 12 in colors too. As in the rainbow. It has five colors clearly visible, but there are more colors on both ends that are not visible to the naked eye. And there is a sound to match each color. See?

There are all sorts of things to try to find and create something new and useful and beautiful.

Your body has a number seven in it, too. There Are seven point or charades beginning from your head and going down the front of your body. Each one connected to certain parts of your body to keep you healthy. There are five more points or Chakras hardly ever noticed in your feet and legs and in your aura above your head.

There is a method being taught all over the world now that touches these points in healing you of stress ailments and others. The method is called by a Japanese name but it is Tibetan: Reiki, pronounced Ray-kee. It is supposed to be taught freely, but in modern times, it is usually not free.

Every child should have lots of melodic music in its life, even before birth. A woman I knew decided to see what would happen if she played very classical Music every day on her tape recorder or phonograph (before many modern things were invented). Her son appreciated that kind of music more than any other and eventually became head of the music department in a university, and directed a famous symphony. So, if you seem to

care for only loud, chanting type music that is what your parents played while you were in the womb. it does make a difference.

There is a lot of scary stuff in Revelations but the verses about the Great Mother are true, as is the description of El Dio LasDiosas and his two groups of 12 Angels-24 of them. Both Dio and Dia have two sets of 12 "Angels" or helpers.

Half are in the third dimension you call "heaven", a half are down here on earth helping rescue people, especially children who can't help themselves. You may have had such a helper keep you from falling off of something, or into something, such as a river. Ella knows she had to have guardian angels caring for her when she was a preschooler and climbed everything, and crossed the river under the moss-covered bridge, hand over hand and did not fall into the water and drown.

Right now, there are dozens of books about Angels on the market. The best ones come only by mail from the publishers of GUIDEPOST MAGAZINE in Camel, New York.

If you are only interested in the world of business, the social, the political, the sports, and entertainment, you are locking yourself out of the Third Dimension. If your interest is in what is being given to you not in what you are giving to help others. That is one of the reversal rules that are often hard to learn.

It is like the rule of gravity, which automatically pull you down, if you don't fight it. The way to raise your inner self is by giving out, not taking in, in every field. There was a French monk who was assigned to polishing the huge pots and pans in a monastery. There were no aids like those available now, so it was very hard work. Yet he did it three times a day and kept his Inner self cheerful and helpful. Eventually, even the king came to him for advice and wisdom.

Libraries and churches have pamphlets about him, Brother Lawrence. All he ever did was work in the kitchen, yet he became so famous that people lined up just to talk with him. Another pattern to notice is that of the medical world. If you have read the ancient stories of Hermes and Mercury (two names for the same being) there was always a caduceus-- a pattern of two serpentine figures weaving up to a pair of wings. You still see it in medical places and writings. Now you recognize it as DNA – only recognized and talked about the past few years. And how important it is just being discovered. So, you see, men like Mercury already knew about the DNA thousands of years ago. When Ella was on the Other Side of the Light, she met him. He had a little wand, he is small too, like a child. With a wand in his hand. The wand has a Lavender light at the end. He gets to hit you with it if you give a wrong answer there.

Have you noticed that all the information so far pertains only to the northern hemisphere of the Earth? Where you are, if you are in North America, Europe, or Russia. Nothing about the southern Hemisphere were South America, most of Africa, Australia, New Zealand, etc. are. Do you know that water runs the opposite

direction there? The seasons are the reverse of those in the Northern Hemisphere? Special colors are in reverse of what is familiar to you?

In the European world over the ages, royal blue

And purple were the most important colors for royalty. Among the Asian lands such as China, Tibet, and Japan, yellow and orange are the colors of the robes worn by the priests. And black and white, white for mourning and black for weddings.

Now we will talk about the stars and the pictures in them, not in the field of Astrology but in Astronomy. They used to be the same thing but telescopes like the Hubble have made a great difference. There are books that tell of the legends and naming of many stars. Very interesting to consult when looking for particular stars or groups of stars. That can be a very interesting area to shop in.

One of the reasons Caesar was named as a god was because of activities in the sky, such as a comet and the lining up of other stars in the inner circle around the sun. But the star actions actually marked the time (a

new age) and birth of a humble man who 300 years Later was declared to be the only son of God. Jesus in English, Issa in other lands.

First, we show you the star map in a college textbook from 40 years ago. It was published by Jacob

Bronowski and Julian Huxley and two other men. It is a beautifully clear map, but note how jumbled the pictures are within it. Running all over the place.

Now what you are going to see or hear about is What is really up there. The same stars form four large pictures, l2 monthly. This is the way they were known thousands of years ago.

In your local libraries, you can find star maps like the one seen on the screen. Make four larger copies, then four more of the 12 monthly ones. There are four nameless days, one between each group of three months. Those nameless ones are for celebration of the Seasons; Spring, Summer, Autumn, and Winter. The fact that the El Dio LasDiosas mind could conjure up so many pictures in the stars, to help you learn about yourselves, makes us respect HIM more than ever.

There is a reason that interesting patterns are in everything THEY created, meaning the whole group of heavenly souls whose ideas were used. First locate ORION, the largest figure on the male side of the map, which you find by folding your map in two and cutting it down the middle.

Now you put a pin hole in every star on both maps, DIO DIA's. Use the male side first, marking where ORION is, then turn the map over to the blank side, put it up to the sun, using a window, or else one of the boxes with a bright light you use for looking at slides.

Now you will begin to see the figure of Orion, riding an elephant which is rearing its front legs Because there is a mouse in its path. Have you heard that an elephant is afraid of a mouse, even though they are of the same family-Hyrax?

Beside Orion on the elephant are two greyhound dogs running. One looks like the shadow of the other.

They are right over the Dog Star. All of those stars had names in ancient Egypt, as did OSIRIS, husband of the goddess Isis. (You went to him when you died, they said. That was 4,000 years ago)

Now you put pin holes in the other half of the star map. It is the Great Mother, who is pregnant. She stands beside La Issa, on which sits a cat. Ahead of A little higher is a Great Dane dog, their protector. This picture shows you why Jesus rode the colt of a donkey into Jerusalem a week before his crucifixion. It also tells us why the Hebrew people lauded him with palm branches. They

knew the star picture and what the donkey represented – the humblest of great teachers.

Now you are going to see another picture in the same stars. This time you need another copy of the Great

Mother star map, holes and all. Now turn it over to the blank side and UPSIDE DOWN to the picture of the donkey and Mother. The lights show through the holes, this time in the form a dolphin leaping through the waves. The DELPHI. The dolphin is still famous for helping anybody who falls in to deep water-- actually or symbolically.

Now you take the fourth copy of the Orion map, put holes in it and turn it over, UPSIDE DOWN to the elephant and Orion scene. Now you have a pheasant cock, the real Thunderbird, flying up from the grasses. Its peacock tail is spread beautifully. He likes a dozen or so hens, while the dolphin prefers just one mate. The Thunderbird name and dolphin are still within your world today as symbols of popular sports teams.

In the years-- thousands of years ago, when Artemis and Apollo were called gods, there was a cave-like place that at first Artemis and a group of women used for healing rites. Later, Apollos priests took it over to learn what the future held for important people and the country, but the men who now held the DELPHI did not have the same

psychic power that the women did, so eventually the center closed.

Apollo and Artemis were twins born of a virgin mother, the legend says, and were the children of the chief god Zeus. The mother kept her children hidden until they were grown so nobody could kill them since They were meant to be important gods eventually.

What is interesting about them is the reversal of traits. Artemis and Roman Diana, were said to be tomboys, while Apollo was a gentle fellow almost Christ-like.

Mary and Joseph did not hide Jesus in Palestine.

But took him to Egypt, where they lived near Cairo

until the boy was four years old. Then Herod the Great, ruler of Palestine, died and his former law about killing newborn babies was no longer a danger to Jesus. Or Thomas, who was born about the same time as Jesus, and also taken to Egypt where he grew up. And he was always considered to be the possible messiah- messenger.

Mary and Joseph took Jesus back to Palestine and there he lived until he was 13 or 14, and then, because many fathers wanted him to marry a daughter of theirs, he left his homeland with the silk merchants whom he knew because they camped near his home in Nazareth Galileo area. He wanted to visit other lands, learn other languages, see what the people believed, and what their

moral creeds were. And eventually came to Tibet, a land hardly heard of or visited. Just like it has been in modern times.

He was called Issa in other lands and Joshua Joseph at home. He stayed about two years in each land he visited, and left when it became dangerous to stay, because he was such a popular speaker, and healer, so full of knowledge that rulers were afraid the people would revolt to make him king.

There is a video about the lands where he went, a map of the whole trip. With the Tibetan history of Issa's life, his teaching, and his death not because the Hebrews turned against him but because the Roman government killed anybody who might lead a rebellion against Rome.

Now, the Bible was put together more than 300 years after Jesus' crucifixion. Compiled from materials FOLLOWERS of the various apostles in different

Lands, remembered for their teachings. Yet, three hundred years earlier, Thomas went to Egypt and formed the COPTIC CHRISTIAN CHURCH, which still exists in Egypt.

What Thomas taught is not in the New Testament, for he did not come back to Rome to help make the book about Jesus. He went to India and other places, living

so much like Jesus did, many people thought he must be Jesus himself. There is still an active St. Thomas church in India, founded by his followers, 2000 years ago.

There is a video now and a book about Jesus's journey with the book written by Janet Bach, wife of the film producer, Richard Bach, who read the book

written by a Russian traveler in 1880. Who fell from his donkey on a path in the mountains of Tibet and was taken by his Sherpa guides to a Buddhist monastery.

He was there three months and there, was shown the Tibetan history of Jesus. And that was the first written in India, where he also studied. The Russian Nicholas Notovich came back home and published the book in Germany. It is now available in English.

"The Lost Years" or "The Jesus Mystery" Among the things Ella learned about the Tibetan lamas, such as the famous Dali Lama, who visits the United States several times a year now that he and his countrymen are refugees in India in the area where India and Pakistan are quarreling over the land, was that, as each one dies, he is reincarnating in North America and Europe.

A book called Reborn in the West talks about modern Lamas, who are building monasteries and living average daily lives. Yet know who they are and are strict with themselves.

This book was written by Vicky MacKenzie and published by the Marlow Company of New York City Another recommend book was written by Elaine

Pagels, one of the translators of the Dead Sea Scrolls, and the book is called GNOSTIC GOSPELS, which quotes from many writers two thousand years ago.

Most of whom were not included in the Bible. Even 300 years after Jesus lived, everything written had to be okayed by the Roman government. So, they picked the Jewish priests as the villains who condemned Jesus to death, rather than their own rulers. The Tibetan/Indian. Record tells what happened and how both countries mourned Jesus's crucifixion. The Jews did not condemn Jesus and what they said is written there. Read the recommended books, see the videos, and other films, and grow a larger view of the rest of the world, then and now.

When Ella returned from the third and longest near-death time. Four weeks off and on at ST. Benedict's Hospital in Ogden Utah, beginning November 12 1960. She began drawing geometric pictures in a language she did not consciously know. After a few months in

Ogden, she returned to California where she had grown up. Meantime, she wrote down the information she has in this message to you, about the atomic element chart, the divisions in it, the Tibetan language chart she made, and the star maps.

It was two or three years later, after she had typed two books for Dr. Hans Von Koerber. A German Baron, who also had read the Notovich book and decided to travel the same path Jesus did, learn the languages and cultures of each area as Jesus did. Then he came to the United States and for 27 years was head of Asiatic Languages at the University of California in Los Angeles. When he retired at age 80, he and his wife built a chalet in the San Diego mountains not far from Where Ella

lived and worked as a reporter for the Escondido TIMES ADVOCATE, 20 miles down the hill. It was there she typed two of his books about languages.

When she asked if he recognized what language she wrote, he did. It was Tibetan. At the time Ella had no idea why she had written in Tibetan, except she knew she had been working in the history of many Cultures on the Other Side of the Light. The geometrical figures formed a Tibetan mandala, which resembles the sand paintings of the American Indians.

Not long after her return to Ramona, California, Tibetan things began to happen to her and she met and married a German Tibetan-Buddhist, who taught her many things about the world she had been in for four weeks, in Ogden Utah.

My dears, few of you have mentally stood back and thought about what you have been taught from childhood on, to hear the unnaturalness of events, or warped version of the truths that can make you feel free and alive.

For one thing, listen to what ETERNAL is telling you. If anything is eternal, it cannot be destroyed, cannot be created again, it has no birth, no death, it travels in a circle. As do the planets, the stars, the weather, the rain and storms, the ocean, the winds. Think about things like that. What can you do to influence any of them?

What can you do to make sure you live ETERNALLY? The ones who tell you that you have to join their group, their teacher, their religion, are Unwittingly "selling you a Brooklyn Bridge." Do you know what that means? It means you are paying somebody for something you already own.

Great teachers tell you that you do live, whether you have a physical ATOMIC ELEMENT body or not.

You shed your mortal body, but nothing can destroy your spiritual one. It belongs to the world of LIGHT, for it is a form of light, it originates as part of a world of light and what you do makes a difference in how bright it is. Very simple. Some people have a malignancy on their spirit, called EGO. It tries to consume, to humble everything or everyone it meets. You cannot affect the weather, the moon and sun's rising and setting, the wind, the rain and storms, but you can affect your inner light. The answer is found in Matthew 25:35-40. A scientific fact.

Anything that is eternal is round like a wheel, like the sun, the moon, the stars. You always had it, you Will always be a part of it, it has no death, only it is part of a glow, part of your eternal self. And it is your THOUGHTS that do have an effect on its glow.

If you know any Bible stories at all, you have probably heard of Sodom and Gomorrah, two places so full of evil, they were destroyed from above – possibly being zapped by tiny asteroids.

The asteroids did not hit those two places by accident. It was because the atmosphere of maliciousness and hurt the THOUGHTS of men

created. Even Lot, who had chosen this pleasant place to live, knew it was full of evil. But he could think of No other family or even one person who was not evil, and he was a pretty weak guy himself. He was willing to sacrifice women of his household to the evil ones, to save the life of the two "angels" who came to warn him and his family to leave.

Soon a great light is to engulf the planet – this planet – whether it destroys it will depend on the THOUGH-ACTION world you create. How much good are you doing for others? Not what you consider good but what Matthew 25:35-40 describes. That is the first question you have to answer when you leave your body of atomic elements for the one of light, which has no atomic elements. It is of another dimension, another world, as real as this one can be with dimensions, LIGHT, warmth, music that heals. So, the most important thing for you to do in self-analyzing, is what Are you THINKING about everybody else?

What do you crave more than anything? Fame? Wealth? Power? Things that can shrivel your soul? That means dimming your soul light to the same condition as that of the trolls in mines of the music, HALL OF THE MOUNTAIN KINGS. You should learn that piece early in life – it is scary-- you will never forget it. You see, they are not dead, they just live a dead life--with almost no light. Same as the MAFIA souls who choose unearned money as their god. Or anybody else who wants to win it all, with only 25 cents in the slot. That is a world out of balance.

You do reincarnate. How soon, where, why, are individual answers for individual people. But some of you really are so good, you may not have to come back to this schoolroom place to try again to learn the lessons it was created to teach you. Some truly worthy souls, like Mother Teresa, come to help those who cannot help themselves. And the whole world calls her a saint.

For those of you who are afraid of death, afraid you will never see people you love, or places again, you should read books like those of Dr. Mooney, who died of pneumonia soon after being drafted into the military service. He didn't know he had died, and his astral body, the light body, traveled across the country to where he hoped to enter a special school. When he got back to the military hospital-morgue, he didn't recognize his body, and found it only because he did recognize his class ring on his finger. There is a video by Dr. Mooney, entitled LIFE AFTER DEATH.

WHO IS ELLA?

Is there any proof that Ella is real? That she has a tremendous knowledge of human nature? How old is she? Where was she born? Did she ever marry, and have children? Does she have a good education? Has she ever taught? Has she traveled to other lands? Has she worked? Does she know about suffering? Healing?

Ella was born April 17, 1915, at the home of her maternal grandparents, Emmanuel and Margaret Groves Jacobsen. In the tiny town of Modena in the Escalante Desert, on the border of Utah and Nevada. Her grandmother was a midwife trained by her mother in-law, Anne Jacobsen Christoferson, of Cedar City, Utah. Emmanuel and his mother were from Denmark, Margaret was from Kanarraville near Cedar City, Utah.

Emmanuel worked in the Silver Reef Mine between Cedar City and St. George, at the time he

found his mother after a seven-year preparation and search. They, he and Margaret, moved to the silver mining town of State line, high above Modena in the 1890s. They had five of their seven children there. They built a mercantile store and hotel and post office there. One building is still left, the only one in the area now.

Ella's father was a young homesteader from Kansas. He met and married Anne Jacobsen when she was 17 and he was 21. He was Jay Moore. Ella was the first of four children. In 1922 the Moore family moved to Southern California, and that was where Ella grew up.

In California Jay Moore built a home in Burbank and that's where Anne, 31, died of a liver ailment. Ella was 13. When Ella was 14, she went to work as a babysitter (nanny) for five dollars a month and room and board,

and put herself through high school, graduating from Alhambra High School, in the San Gabriel Valley. She won a scholarship to Occidental College, but five dollars a month was not enough income to attend school. By working mostly at night in six colleges and universities she eventually earned a degree when she was 66 years old. Over the years, so she could walk, her hips and pelvis have metal rods in the right leg and partial plastic in the pelvis. She had 15 surgeries altogether, two on her ears, so she could hear, four on the eyes, four on the hips, etc. so she could walk and live internally, mostly because of parts of her body calcifying, turning to stone.

During World War II time she was a reporter for the SALT LAKE TRIBUNE. Then for the Ogden STANDARD EXAMINER for nine years, and

published a weekly in the small town she and her husband lived in for 18 years.

With the newspaper, she recruited PTA members to vote for a very good library system in Davis County, plus nursery schools for little children whose mothers worked. She formed Girl Scout troops in Salt Lake City, Davis County, and Ogden, Utah. In addition, she personally led scout troop at the state school for the deaf, then the blind. And at the same time the State Industrial School for Girls.

When someone asked her husband what it was like to be married to a woman so involved with community Services? He said "I tell you this, I've never been bored." Because of her service with the handicapped girls, Ella was one of 30 women from around the world, invited to come the New York in 1948 to compile a new handbook for leaders of handicapped girls. She also went to Washington DC. afterwards.

On January 3, 1949 she was told by specialists that she would never walk again because of the condition of her hips and pelvis. She used a wheelchair for 18 years, most of the time and continued mothering homeless young people.

She had no mate in the 1960s and part of 1970. Twelve young people came to live with her during that time. Seven of them were born in other countries.

In 1966 she received her first talking book for the blind. She could no longer see her news copy, could not walk, and wondered what she could do to continue earning a living? The talking book told her what was possible, she could teach newcomers to the United States how to speak, read, and write English. The book was about the world service Dr. Frank Laubach did in the field of literacy. Ella took her training from Bea Avery of Escondido, San Diego County, tutored a woman from Korea, and two from Mexico for a year.

Then she was invited to come to Mexico and teach, as a volunteer, in the Futuro del Oro project overlooking the Pacific Ocean. She did teach for two years three days a week and then went by bus farther along the border to Tecate to teach children at the Fundacion Para Ninos orphanage, founded by Harry Hold who brought thousands of Korean babies to the United States, because Koreans do not accept children of mixed blood, and these babies were fathered by American servicemen. Mr. Holt died before the Mexican orphanage was completed, but his Mexican woman partner, a secretary to five lawyers', finished the project and filled it with children of all ages-- around 40 at one time, all learning English to be prepared for adoption in the United States.

When Ella went to New York to help outstanding Girl Scout leaders from around the world, she met a woman who prepared her for a totally different service, teaching children with cerebral palsy. The first year Ella was in the wheelchair, she was asked to be a volunteer teacher of helpless babies and little children In Ogden, Utah. Children who had hardly ever been out of their cribs, had never had anything to stimulate their minds – because hardly anyone knew what to do, not even the doctors.

Ella accepted and from a wheelchair, taught the parents, women who volunteered to work with little people who did not open their eyes or respond the first six weeks of

classes, then on a Monday morning all of Them were awake, ready to play with the colored toys, dishes, materials, they had been hearing about and touching for six weeks. It was a joyous occasion.

As a result of that class, soon two rooms for handicapped children and special restroom facilities and a fenced in play yard, to protect these children from rambunctious healthy ones, was built. Soon after wards, hundreds of little ones with the same six types of handicaps were traveling on buses and as they grew up had special training for jobs and education.

The young woman who inspired Ella to do the service with children was a victim of cerebral palsy, considered an idiot until she was 16, then tested and found to be bright and wise. She was trained, put in leg braces, learned how to clasp her hands when speaking-

-using only the half of her throat that was not paralyzed. She graduated from college when she was

24, went to teach in Ohio's institution for severely handicapped children. Her name was Jo Wentworth. Bless her holy name.

Ella was not a pretty little girl. She had dark red hair and freckles from the age of two, in addition to limping and not seeing. When she played softball, she could bat the ball but Atsuko Masaki, her Japanese friend, ran for her. Teachers never mentioned any of her handicaps to Ella, but helped her by treating her like everyone else.

The three near-death experiences began in the summer of 1958. She had her second hip surgery, had been in traction for six weeks, came home and had a

gall-bladder attack so severe, the doctor gave her a shot of morphine and out she went through the top of her head, down the dark tunnel, towards the light. When she went through the little "PING", like the ones in grocery store doors, a strong male voice spoke: "Do you know where you are going? Turn around and go back." She did and when she came back, she told her doctor and her husband: I've been to some place.

They said: "We know". She had been gone for eleven minutes, during which time the doctor used every method he knew to revive her.

The second near death experience was totally different. It was caused by blood sugar dropping too low, though she did not know it. She was ill all day, and finally swooped out of her body, in to her bedroom. She moved in her astral body towards the door, and then saw a line of women, dressed as nuns, but with eyes closed as in sleep, moving towards her. She was frightened by them and exclaimed: "I thought my mother would come for me". As she spoke, she snapped back in to her body, and told her son and husband, who were sitting by her bed, "I need a cup of tea with sugar in it". Her husband immediately made tea and put sugar in it and brought it to her. She drank it and was immediately back to normal.

The third near death experience began on November 12, 1960 at 6 pm. In the morning, the

Guardian's voice told her: "You are going to die at 6 pm.

tonight. You will not be back here so giving everything away to those you want to have them." So, Ella wrote down the message and put it under her pillow so the family would know she knew she was leaving. She told her son and daughter-in-law if they would give their upright piano to the church, which needed a piano for the children's department, she would give them her blond spinet piano. She gave away books and other things too. The day was spent moving out things.

She was left alone late in the day, and at 6 pm. the heart pain increased so much, Ella went up out of her head, just in to her bedroom towards a tall figure of Light. "Are you Jesus?" she asked. "I am if you think I am." was the answer. Then the phone rang. The Being of Light disappeared. Ella floated to the phone on her bed and tried to raise the receiver, but couldn't budge it. When she did not answer, soon her family came home and took her to St. Benedict's Hospital in Ogden. There she stayed in a coma like stage for four of the five weeks she was there. No visitors were allowed.

Sister Mary Margaret, head of St. Benedict's Hospital sat beside Ella's bed the four weeks and did her office work there, plus wrote down messages Ella spoke from wherever she was, spiritually. Later, Sister Mary Margaret asked Ella if she knew who Saint Teresa,

Little Child of Jesus, was. Ella had never heard of her. Then Sister Mary Margaret told Ella about the little French nun, in the same condition as Ella was.

Teresa spoke in French, Ella in English, but the words Were exactly the same. The nun died in 1897, at age 25.

Immediately after St. Teresa died, she began to heal people, and so was named a Saint by the Roman Catholic Church.

In the fall of 1941, when my husband, Ross D. Hyatt, took a mineralogy course on Monday nights at the University of Utah, Salt Lake City, I first saw an atom chart in our text book. I was delighted to find the material about atoms, but distressed that the chart was so incomplete and not in the right order. I was also thrilled with the beautiful pictures of crystals.

It was 20 years later, in June, 1961, that I drew the correct atom chart. It occured this way: I had been taking care of a sister of a good friend, this woman had just had an amputation of her leg. The last day, A Saturday, she gave me $50 for my help. I immediately took a bus to downtown Ogden, Utah, and hurried into a large book store on the main street. I walked down what I felt was the right aisle and hld my hand above the stacks of books the clerks were unpacking. Then on the shelf I felt a magnetism. I put my hand over the stack on the display counter-shelf, the lights turned on inside of my hand and me. The book was a science text by Jacob Brownoski and Julian Huxley and two other men.

I took the book home to my apartment, sat down and looked at the atom chart, still incomplete and immediately saw in my mind the pattern again:

and so drew a sketch of the atom chart. Then I typed what you have on the other side of this paper.

The next day, Sunday, I did the two large star maps and 12 monthly ones.

I always wanted to take a course from Brownoski while he was teaching at the university in the San Diego-LaJolla area but my life was moving me eastward at that time. However, while giving training in the Apple Valley-Lucerne Valley area of the High Desert of California, I heard all the weekly sessions he gave on television, which came on about the time he died.

Another great mind I would have liked to have known was Robert Graves, author of such interesting books. His book, I CLAUSIUS, was such a fascinating story of the Caesars and their contemporaries. He also wrote one about the first (Roman) century time of Herod the Great-- called KING JESUS. In it is quite a bit about the beliefs of the people, which included the importance of the donkey in the stars-- the reason the people honored Jesus when he rode the COLT of the donkey into Jerusalem a week before his crucifixion.

Margaret Hyatt Young

The 1.08 atom group, called Deutarium by people now,
I call Hy-8, on the chart.
Tibetans have 108 beads on their rosary for this reason.
The hyet word was prep-greek for rain.
It is also my name.

Margaret Hyatt Young
Box 404 Parowan, Utah 84761

1 H Hydrogen	18 Ar Argon	19 K Potassium	36 Kr Krypton	37 Rb Rubidium	54 Xe Xenon	55 Cs Cesium	72 Hf Hafnium	73 Ta Tantalum	90 Th Thorium	91 Pa Protactinium	108 Hs Hassium (Hy-8)
2 He Helium	17 Cl Chlorine	20 Ca Calcium	35 Br Bromine	38 Sr Strontium	53 I Iodine	56 Ba Barium	71 Lu Lutetium	74 W Tungsten	89 Ac Actinium	92 U Uranium	107 Bh Bohrium
3 Li Lithium	16 S Sulfur	21 Sc Scandium	34 Se Selenium	39 Y Yttrium	52 Te Tellurium	57 La Lanthanum	70 Yb Ytterbium	75 Re Rhenium	88 Ra Radium	93 Np Neptunium	106 Sg Seaborgium
4 Be Beryllium	15 P Phosphorus	22 Ti Titanium	33 As Arsenic	40 Zr Zirconium	51 Sb Antimony	58 Ce Cerium	69 Tm Thulium	76 Os Osmium	87 Fr Francium	94 Pu Plutonium	105 Db Dubnium
5 B Boron	14 Si Silicon	23 V Vanadium	32 Ge Germanium	41 Nb Niobium	50 Sn Tin	59 Pr Praseodymium	68 Er Erbium	77 Ir Iridium	86 Rn Radon	95 Am Americium	104 Rf Rutherfordium
6 C Carbon	13 Al Aluminum	24 Cr Chromium	31 Ga Gallium	42 Mo Molybdenum	49 In Indium	60 Nd Neodymium	67 Ho Holmium	78 Pt Platinum	85 At Astatine	96 Cm Curium	103 Lr Lawrencium
7 N Nitrogen	12 Mg Magnesium	25 Mn Manganese	30 Zn Zinc	43 Tc Technetium	48 Cd Cadmium	61 Pm Promethium	66 Dy Dysprosium	79 Au Gold	84 Po Polonium	97 Br Berkelium	102 No Nobelium
8 O Oxygen	11 Na Sodium	26 Fe Iron	29 Cu Copper	44 Ru Ruthenium	47 Ag Silver	62 Sm Samarium	65 Tb Terbium	80 Hg Mercury	83 Bi Bismuth	98 Cf Californium	101 Md Mendelevium
9 F Fluorine	10 Ne Neon	27 Co Cobalt	28 Ni Nickel	45 Rh Rhodium	46 Pd Palladium	63 Eu Europium	64 Gd Gadolinium	81 Tl Thallium	82 Pb Lead	99 Es Einsteinium	100 Fm Fermium

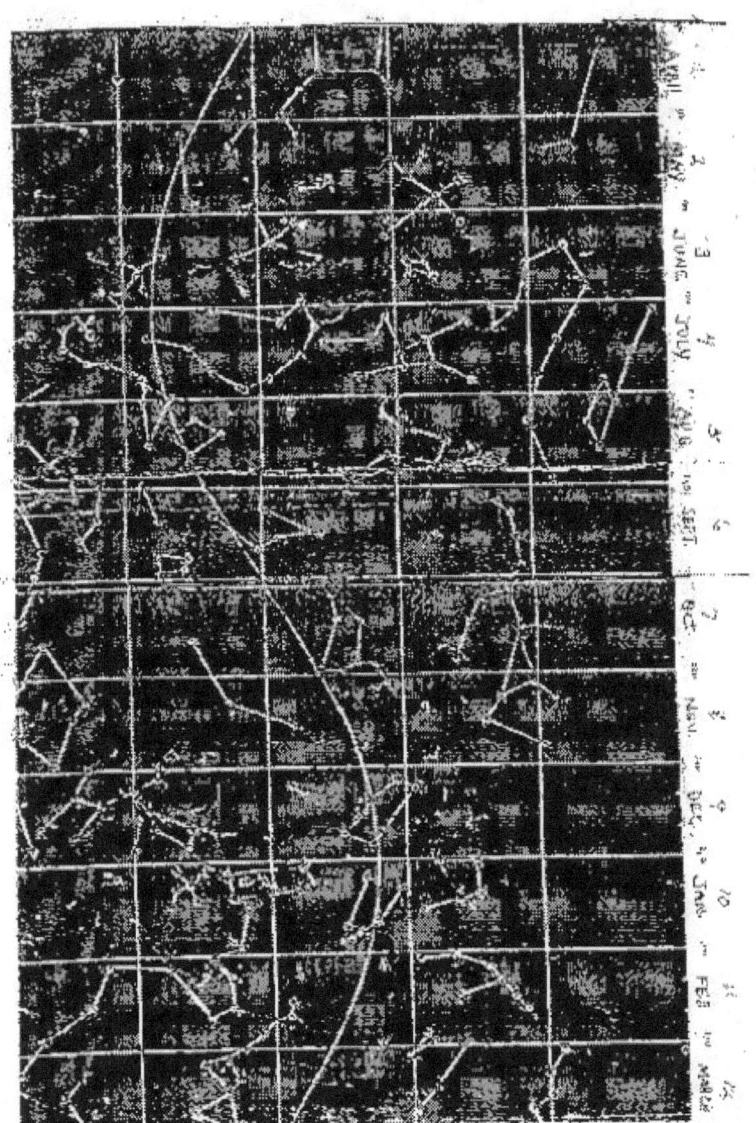

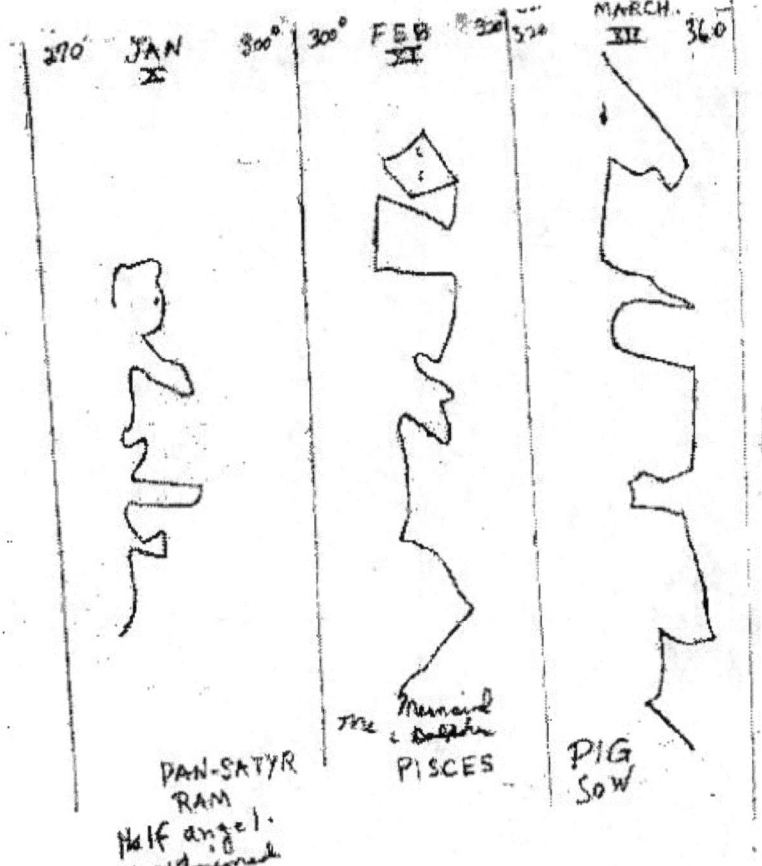

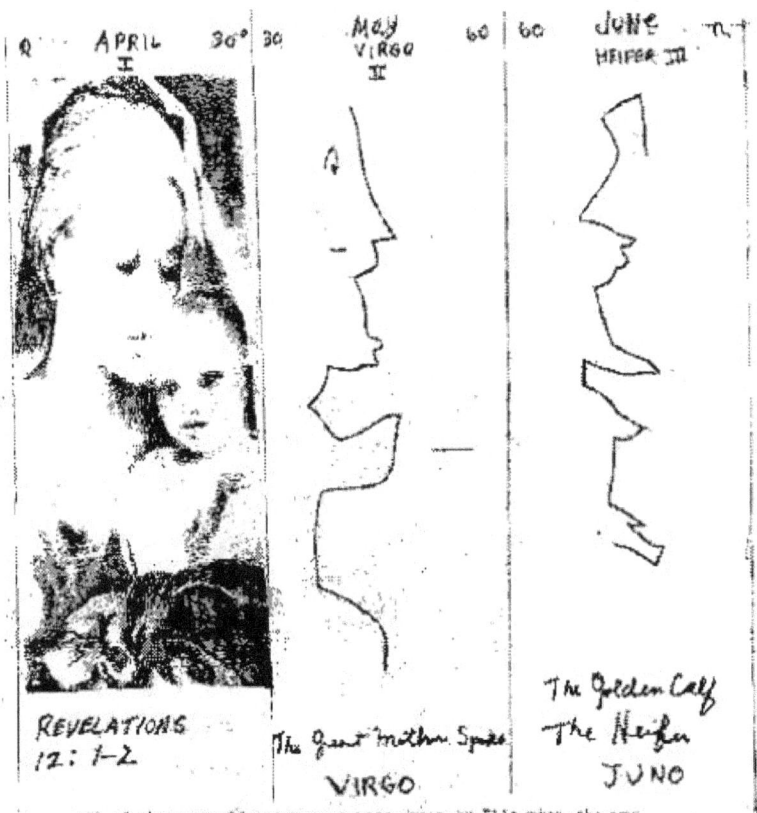

All of these monthly star maps were drawn by Ella when she was four years old, except that of the GREAT MOTHER, in April. This time she used a picture of a BLACK mother, in Africa, the area where humans first developed. El Dio told his daughter what to draw and these pictures are what she did. Some of your lands still have some or even most of these star pictures, such as CHINA does with the pig as the 12th star. Library books have all this material and names used in different lands and cultures.

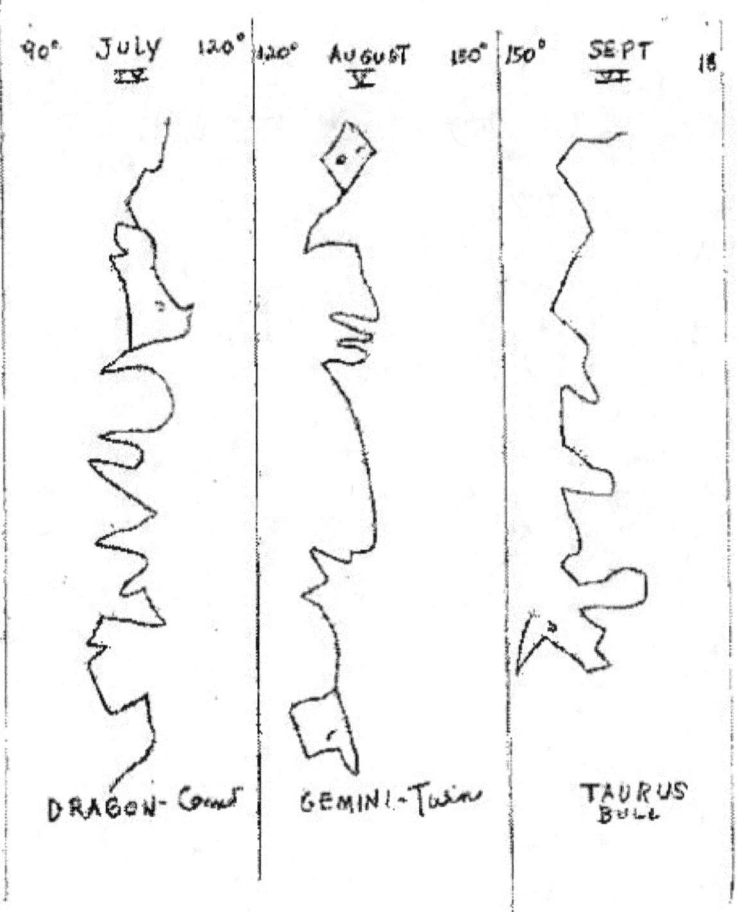

Oct. XII	Nov. VIII	2.40 DECEM. 27 IX
HORUS HAWK or NEGUS	HIPPO (prius) Renverser	CAM-BAR CAMEL Bear

The Tibetan Vowel Chart — 1961

Tibetan

A	E	I	O	U	O	I	E	A
A A A				V				A A A
SA	BE	CI	DO	UV	OT	IX	EW	AZ
AF	EB	IC	OD	VU	TO	XI	WE	PA
FE	BI	CO	DU	YV	UT	OX	IW	EP
EG	IB	OC	UD	VY	TU	XO	WI	JE
GI	BO	CU	DY	AV	YT	UX	OW	IJ
IH	OB	UC	TD	VA	TY	XU	WO	NI
HO	BU	CY	DA	EV	AT	YX	UW	OK
OK	YB	YC	AD	VE	TA	XY	WU	PO
KU	BY	CA	OE	IV	ET	AX	YW	UP
UB	YB	AC	ED	VI	TE	XA	WY	MU
LY	BA	CE	DI	OV	IT	EX	AW	YM
YR	AB	EC	ID	YO	TI	XE	WA	RY

				Y				
Y	A	E	I	O	I	E	A	U
1	2	3	4	5	4	3	2	1
108	107	106	105	104	103	102	101	100
9	8	7	6	5	4	3	2	1

This is the TIBETAN vowel sound chart written by ELLA in October 1961 in San Jose, California where she had gone to work for the USO, in the office on the 6th floor of the Commercial Bldg. She chaperoned a bus load of teenage girls on Friday and Saturday nights to Fort Ord and Monterey to dance with young soldiers bound for Vietnam.

www.ingramcontent.com/pod-product-compliance
Lightning Source LLC
Chambersburg PA
CBHW080955220526
45465CB00008BA/3294